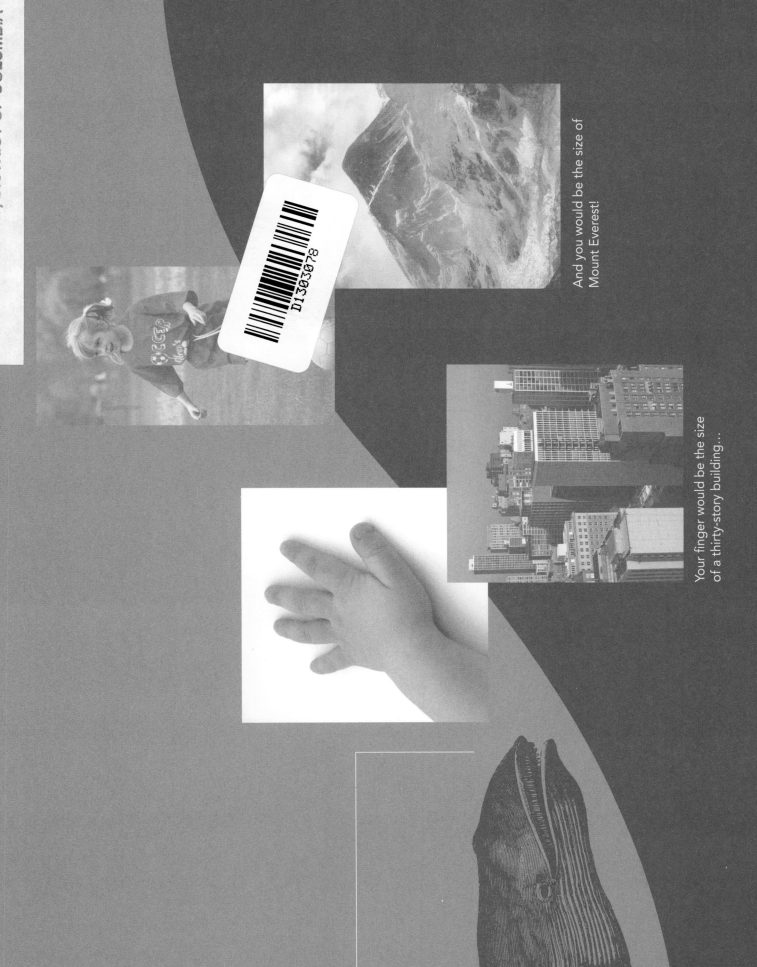

And you would be the size of
Mount Everest!

Your finger would be the size
of a thirty-story building…

The Invisible ABCs

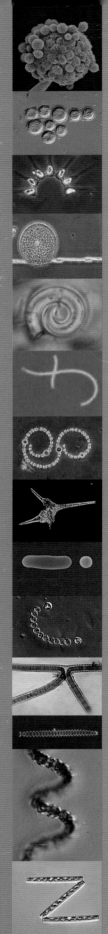

Dedicated to Ginny Hempleman and her many elementary school students, and
to those who love to spend time with curious and enthusiastic children.

I would especially like to acknowledge Jeff Holtmeier, Director, ASM Press, for his encouragement
and support of a children's book about the benefits of microorganisms. Ellie Tupper (ASM Press) and
Debra Naylor and Josh Rubinstein (Naylor Design Inc.) have taken a simple idea and made it spectacular.
In addition, I would like to acknowledge the many photomicroscopists who were kind enough to allow
their work to be used here: Robert Kistler, Morgan Vis, Donald Ferry, Dennis Kunkel, D. J. Patterson,
Wim van Edmond, Petr Znachor, and John Priscu.

Learn more about the microbes at http://www.TheInvisibleABCs.org

Copyright © 2006 ASM Press
American Society for Microbiology
1752 N Street, N.W.
Washington, DC 20036-2804

Library of Congress Cataloging-in-Publication Data

Anderson, Rodney P.
  The invisible ABCs by / Rodney P. Anderson.
        p. cm.
  ISBN-13: 978-1-55581-386-4
  ISBN-10: 1-55581-386-0
  1. Microorganisms—Juvenile literature. 2. Microbiology—Juvenile literature.
  3. English language—Alphabet. I. Title.

  QR57.A53 2007
  616.9'041—dc22

                                          2006016975

Send orders to:
ASM Press, P.O. Box 605, Herndon, VA 20172, U.S.A.
Phone: 800-546-2416; 703-661-1593
Fax: 703-661-1501
Email: Books@asmusa.org
Online: estore.asm.org

The Invisible ABCs

Exploring the world of microbes…

Rodney P. Anderson

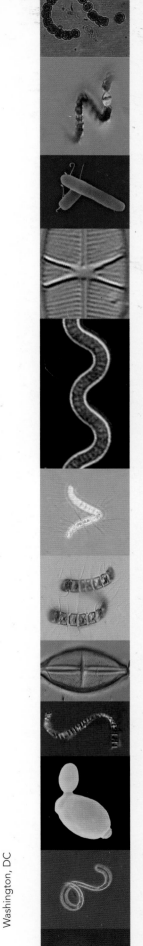

ASM PRESS

Washington, DC

here is an unknown world too tiny to see.

It is a hidden land where monsters with oozing feet catch and eat their smaller neighbors.

It is an invisible ocean where floating grasslands surround those who calmly feed in the sea of green.

This unseen place has filled and changed the Earth since life began.

You and I cannot survive without these strange creatures.

Amoeba

Protozoan

Staphylococcus aureus
on human hair

They live all around us—and on us—and in us.

They keep us healthy, provide us with food, and even help make the air we breathe.

Who lives in this world of the hidden?

It is microorganisms—living things that are too small to see.

Discover the unknown and see the unseen as we explore the ABCs of the microbes we can't live without.

Anabaena

Vortecella

A is for Algae

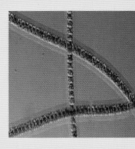

Algae make up the invisible grasslands of our unseen world. They live anywhere that is wet. Algae can look like colorful jewels, pieces of green-colored hair, or glass balls floating in space.

Algae are food for tiny microbes and big animals. They are at the beginning of a food chain where small creatures are eaten by bigger creatures that are eaten by even bigger creatures. The fish that we eat have eaten smaller fish that have eaten algae.

Like trees and grass, algae need water and sunlight to grow. As algae grow, they make oxygen—an important part of the air we breathe.

a

Would our world be the same without algae?

Cymbellan

Pleodorina

Staurastrum

Closterium

Zygnema

Green algae

Algae live in swamps (top) and lakes (above)

B is for Bacteria

Bacteria are too small for us to see without using a special instrument called a microscope, a tool that makes small things look big.

Bacteria come in different shapes and sizes. They live everywhere—in water, dirt, and air and in and on plants and animals. There are billions of bacteria living in and on you right now: floating invisibly in your eyes, hidden between your teeth, growing on your skin, and eating the leftover food inside you. Most bacteria are friendly and help keep us from getting sick from the disease-causing microbes.

How does it make you feel to have all these friendly microbes with you every day?

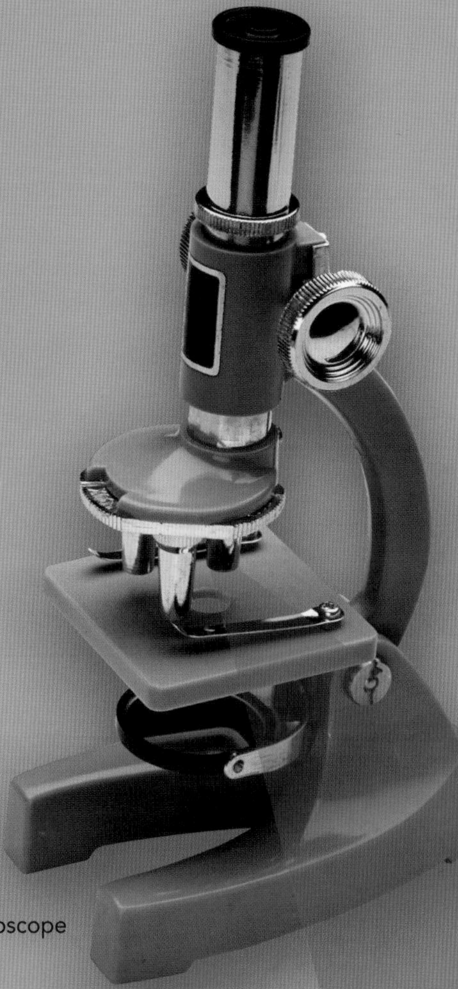

Staphylococcus aureus

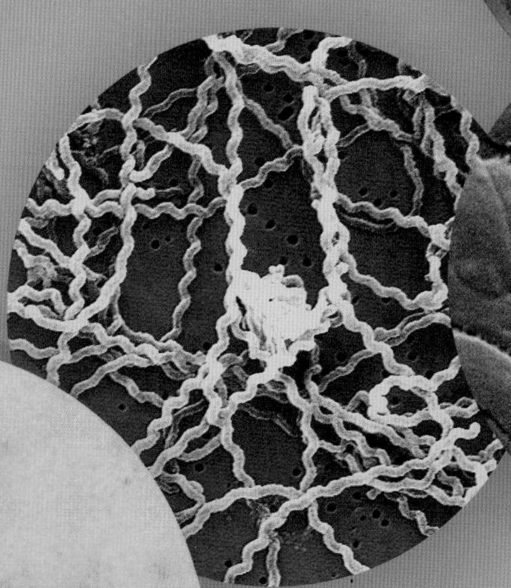

Vibrio

Leptospira

Spirillum volutans

Decaying tree

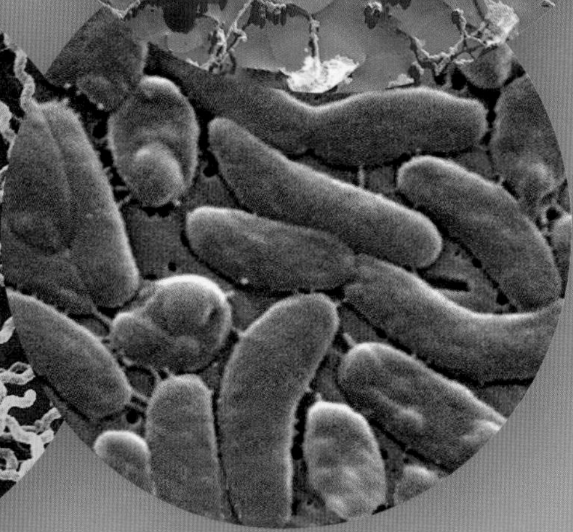

Microscope

Bighorn sheep

C is for Cows

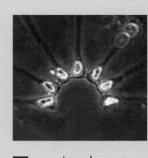

Do you know why cows can eat grass and we can't?
It's not because grass tastes bad!

It's because plants like grass are mostly made of cellulose, a food that human stomachs can't process. Cow stomachs are filled with billions and billions of special bacteria that we don't have. The cow gives all those bacteria a warm place to live and lots of grass to eat. The bacteria break down the cellulose into food for themselves and leave behind nutrients that help cows grow.

Most animals that live on grass and leaves have these bacteria—how many can you name?

Studying digestion of stuff from a cow's stomach

Above and below:
Bacteria from a
cow's stomach

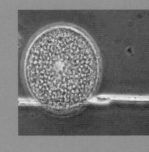

is for Diatoms

Diatoms are a kind of algae that make their own houses to live in—tiny glass-like boxes that look like colored jewels in different shapes and sizes. Like all algae, diatoms live in the water and use sunlight to make food. They release the important oxygen we need to breathe.

What would it be like to be a diatom and live in a glass house under the water?

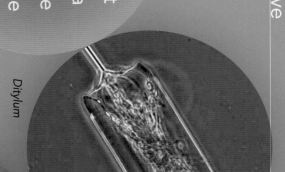

Ditylum

Actinoptychus

Marine diatom fossil

Fossil diatoms
from Barbados

Pinnularia

Glacier

Black smoker

Decaying tree

Cave

Coscinodiscus

Achnanthes

is for Everywhere

There are many kinds of invisible microbes—bacteria, algae, molds, yeasts, viruses, and protozoa. Microbes are everywhere!

They float in clouds in the sky. They live in the deep, dark oceans where they eat the minerals underwater volcanoes spit out. They live frozen in ice where nothing else can live. Even if you dug a hole very deep into the earth, you would find microbes.

Coral reef

If we can find microbes everywhere on the Earth, do you think they may live on other planets?

Licmophora flabellata

Periphaena decora

F is for Food

Microbes help make foods we like to eat. Yeast puffs up bread so that it is fluffy and soft. Bacteria produce gas that makes holes in Swiss cheese, and they make yogurt thick and tasty. They even help in processing cocoa beans for chocolate! Bacteria and yeasts are two kinds of microbes that we can use to make food that's good to eat.

Do you know of other foods that are made with the help of yeast and bacteria?

Propionibacterium

Cocoa beans

Swiss cheese

Yeast cells

Yogurt and cereal

Lactobacillus

G is for Gas

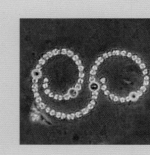
Enterococcus

Some bacteria that live inside us make gas—not the gas we put in our cars or the gas that is used to heat our homes—but the smelly, embarrassing kind. What food do you think these bacteria eat? Beans!

When beans move through our intestines, some of the leftover stuff is used as food for the gas-making bacteria. In healthy people, bacteria that make poop smell aren't bad. Some of these bacteria make important vitamins that our bodies need to grow and stay healthy.

What other foods do you think gas-making bacteria like to eat?

Beans

Three types of bacteria

Escherichia coli

h is for Healthy

Microbes live in and on us and help us to be healthy. We know that microbes inside us make vitamins that our bodies need. Our body's normal population of microbes doesn't leave many nutrients for the microbes that make us sick.

Who else do you know who helps you stay healthy?

Vitamins

Staphylococcus epidermidis

Staphylococcus on a human hair

Lactobacillus

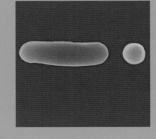

is for Immunity

Immunity means that our bodies can fight off disease-causing microbes. Immunity keeps us from getting sick. You can get immunity by having a disease or by getting a shot.

Have you ever had the measles? How about your parents or grandparents?

Before shots were invented, almost all children got measles. Now many children never get measles, mumps, or whooping cough because their shots have built immunity to these diseases.

Measles virus

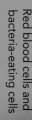

Red blood cells and bacteria-eating cells

"Rubella Fighter" after his shot

Would you rather have a shot or be very sick with a preventable disease?

Bacteria-eating blood cell

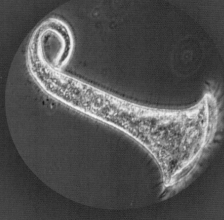

is for Jenner

Do you like pus-filled sores on cows? Dr. Jenner didn't either until he realized that cow pus could be used to provide immunity against the deadly disease smallpox. Many years ago, the virus that causes smallpox made people very sick, and many died.

At first people laughed at Dr. Jenner and even thought that getting a shot would turn them into cows! However, Dr. Jenner was right: shots do prevent us from getting all sorts of diseases. Isn't it weird that we are healthier today because of sick cows and a scientist who learned about immunity?

Do you think you would turn into a cow if you got vaccinated?

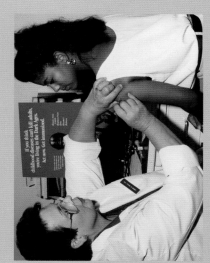
Dr. Edward Jenner

Modern vaccination

Smallpox virus

Old drawing of cowpox sores

The Cow-Pock — or — the Wonderful Effects of the New Inoculation! — vide the Publication of J Anti Vaccine Society

A cartoon from 1802 implying that vaccination turned people into cows

is for Kaleidoscope

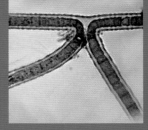

Have you ever seen the changing colors and shapes inside a kaleidoscope?

A kaleidoscope has mirrors inside that help to make new pictures every time we look. The invisible world of microbes is like a kaleidoscope—it is filled with different shapes and colors that we don't see until we look inside.

Only a small part of this world has been discovered. Every day new microbes can be found.

Diatoms arranged in a rosette

Nauplii

Arranged diatoms

Globigerina

Bacillaria paxillifer

Would you have fun exploring the unknown world of microbes?

L is for Lichens

The microbes that make up lichens must live well together. A lichen can't survive unless the different microbes help each other.

Algae—the green microbes that make food using sunlight—live inside the lichens.

The outside of a lichen is mold. The mold eats the algae's leftover food and protects the algae in return.

Microscopic view of the inside of a lichen

Lichen on tree in Northeast Boreal Forest

Do you have a friend you wouldn't want to live without?

Lichens growing on rocks

Arthroderma

M is for Mold

Molds are fuzzy, hairy microbes that grow wherever there is enough water and nutrients, like a wet forest, a piece of fruit, or even a musty bathroom. Moms think mold in the bathroom is disgusting and try to scrub it away. Molds sometimes look gross, but we can be thankful for molds.

Some molds make medicines we call antibiotics. When children are sick from disease-causing bacteria, doctors give them an antibiotic to help make them healthy again.

Mucor

Hypocrea

Moldy grapefruit

Has a doctor ever given you an antibiotic for a sore throat or an ear infection to help you feel better?

Antibiotic pills

Botrytis

Mold growing on a wall

Mold colonies in petri dishes

N is for Nutrients

What happens to the leaves after they fall off the trees before winter? Microbes eat them!

Dead plants and animals are what microbes in the dirt eat. They leave behind important nutrients that plants need to grow. When we see green grass, flowers, and leafy trees, it's because microbes are doing their job.

Leaf litter

Pseudomonas

Nitrosomonas

How would your neighborhood change if microbes didn't eat dead leaves and grass?

Sunflower with bee extracting pollen

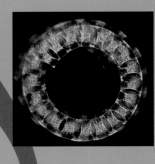

is for Oxygen

We must have oxygen to live. We use oxygen with every breath we take. Why don't we run out of oxygen?

Tiny green algae that live in the oceans, lakes, and rivers each make a little bit of oxygen every day. Working together, algae provide most of the oxygen for all the people and animals on Earth.

Green shows algae in the oceans and plants on land

Grass and trees make oxygen also

Stentor with *Chlorella* algae living inside

Euglena

Pediastrum and *Coelastrum*

Eudorina elegans

What would happen if algae couldn't live in the oceans?

Lead tree, *Leucaena leucocephala*

Amoeba

Ciliate

P is for Protozoa

In the world of microbes, protozoa are like the deer and wolves of our world. Many protozoa eat the algae that make their food from the Sun's energy, the same way that deer eat grass that makes its food from the Sun's energy.

Wolves eat the deer for food just like some other protozoa eat the microbes that feed on algae.

Stylonychia

Vortecella

Diophrys

What would happen to wolves if they couldn't eat deer, or to deer if they couldn't eat grass?

q is for Questions

Microbiologist using a microscope

Asking questions is one of the ways we learn. When we ask questions about the microbial world, many times no one knows the answer. The microbial world is mostly undiscovered. How do microbes keep us healthy and how can we stop them from making us sick? Can we find more microbes that make medicines? How can we use microbes to make our world better?

Would you enjoy becoming a person who finds the answers to questions about microbes?

Anabaena

Gallonella

Blood with cells that eat microbes

Ophiocytium

Rr is for Roots

Plants need roots to get water and nutrients from the soil. Some plants, like those in the bean and pea family, get help from their microbial friends. Bacteria live inside the roots of these plants and make nutrients that the plants need to grow. In return, bacteria get a nice place to live and can eat some of the sugar that the plants make.

What would happen to the plants if bacteria didn't help them?

Peas and beans

Bradyrhizobium

Roots with nodules

A week in the life of a pea

Soybeans

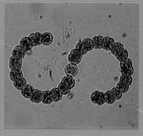

is for Sewage

Sewage is what gets flushed down the toilet. Have you ever wondered what happens to all that poop? It becomes food for the microbes that are grown at sewage treatment plants. It's a feast!

After they eat all the poop, the microbes are killed and the leftover water is then safe to send back into the environment.

Aren't we lucky that microbes help clean our dirty water?

Sewage treatment plant

Zoogloea

Sewage treatment plant

Ciliate

Amoeba

Ciliate

Protozoan

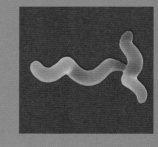

is for Termites

Termites are insects that eat wood all day long. But without help, most termites can't use the wood they eat for food. They need microbes! Like cows, termites have helpful microbes in their stomachs that break down the cellulose in wood into the food the termites need. The termites live on the microbes' leftovers.

t

Termites eating wood

Termite damage

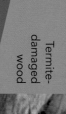

Termite-damaged wood

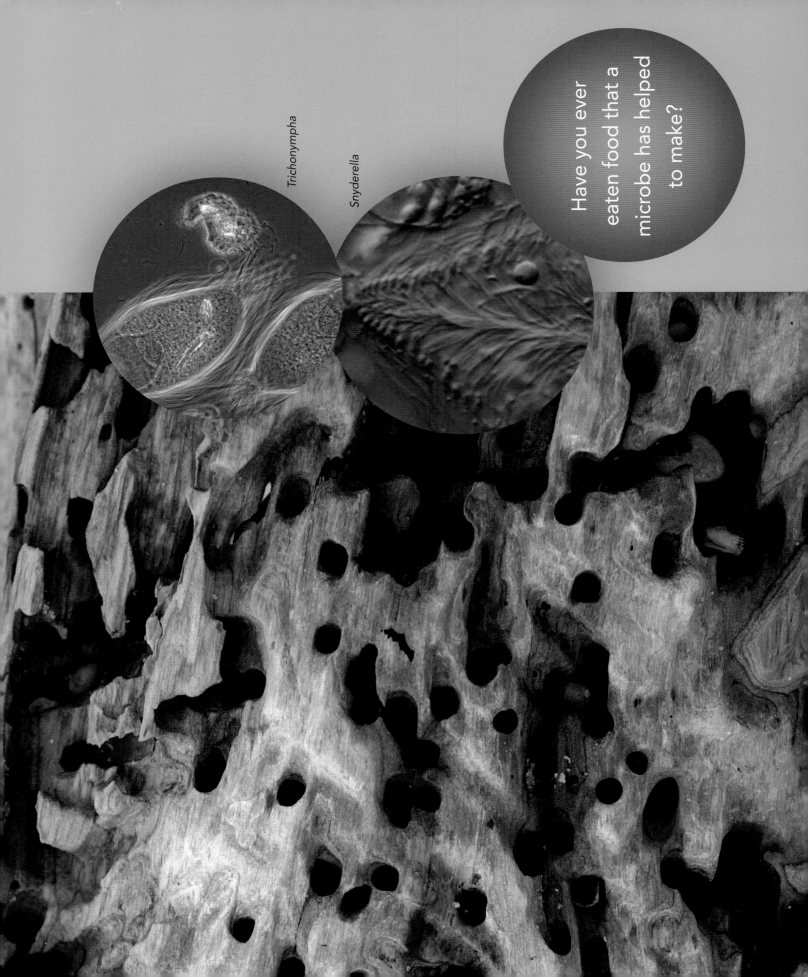

Trichonympha

Snyderella

Have you ever eaten food that a microbe has helped to make?

is for Useful

Microbes are the most useful of all living things. Microbes help to make foods such as cheese, yogurt, bread, and chocolate. Microbes help clean up pollution from our environment, and they're used to help remove copper and gold from the rocks that miners dig up. Some microbes make special fuel for our cars and trucks, and others make medicines.

Can you
think of other
useful things
microbes do?

Oiled
seabird

Making cheese

Antibiotics

CAUTION: ...
RX 33 X 338646
TAKE (MAKE 1 CAPS...
UNTIL STARTING ON...

LINDAMYCIN...
PENICILIN 21
QTY 28 EFILLS LEFT
REFILLS

V is for Viruses

Viruses are the smallest of all microbes. They are bad news. They stick to the insides of your nose, mouth, lungs, skin, or intestines. They can give you a cold, the flu, chicken pox, warts, diarrhea, a sore throat, or even a more serious disease. These viruses are not our friends. How can we fight off these invisible enemies?

Wash your hands well after you use the bathroom, cover your mouth when you cough and sneeze, and be sure to get the shots that keep you from getting sick.

Rotavirus

Microbes released from a sneeze, growing in a petri dish

What a sneeze looks like

Ebola virus

Left: Influenza virus

Have you ever had a virus make you sick with a cold or the flu?

is for Whales

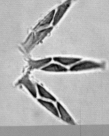

Whales are the largest animals on earth. What do giant whales have to do with tiny microbes, the smallest of living things? Some of the biggest creatures on earth feed on the smallest. Certain whales have special bony plates in their mouths. They open their mouths as they swim, and the plates capture billions of algae, protozoa, and tiny animals called zooplankton drifting in the water. With so many microbes in the oceans, the whales that eat them can grow to be the giants of the world.

Besides being eaten for food, why are microbes in the oceans important?

Gray whale

Copepod

Starfish larvae

Calanoid copepod

Lithodesmium intricatum

Achnanthes

Noctiluca scintillans

is for eXtreme

Some places on this planet are extreme. That means that they are too hot, too cold, or too salty or acidic for people, plants, and animals to live. But some microbes love these places that would easily cook or freeze us. Some bacteria can live in water boiling from the heat of geysers or volcanoes on the ocean floor. Others survive in rivers of ice called glaciers that have been frozen for thousands of years. There is a big lake called the Dead Sea that is so salty that fish, animals, bugs, and birds can't even live there. But microbes can!

Geysers and glaciers are spectacular places to visit, but could you live there?

Black smoker

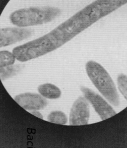

Bacillus infernos

Ice bacteria

Isanotski
Peak volcano,
Unimak Island

Halobacterium

The Dead Sea

Microbial streamers
in a hot spring at
Yellowstone National Park

Strokkur Geyser,
Iceland

Y is for Yeast

Yeast eats sugar and makes alcohol and carbon dioxide gas. We use yeast to make bread. Have you watched bread being made? The dough rises! As the yeast in the dough eats the sugar, it produces gas and the dough puffs up. Yeasts also help to make wine and beer. The yeast eats the sugars from the crushed grapes or malt and leaves behind alcohol.

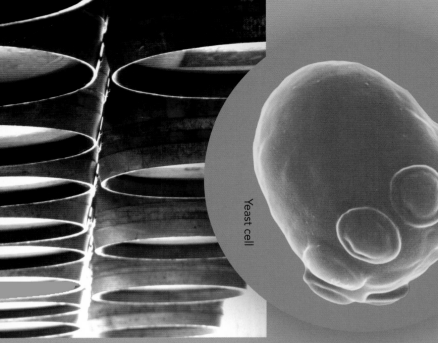

Yeast cell

Brewer's yeast

Wine barrels

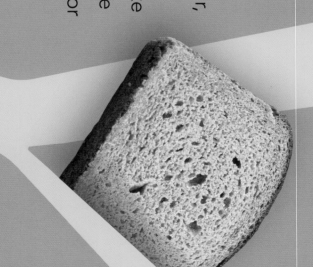

How many foods can you think of that are made using microbes?

Z is for Zoo

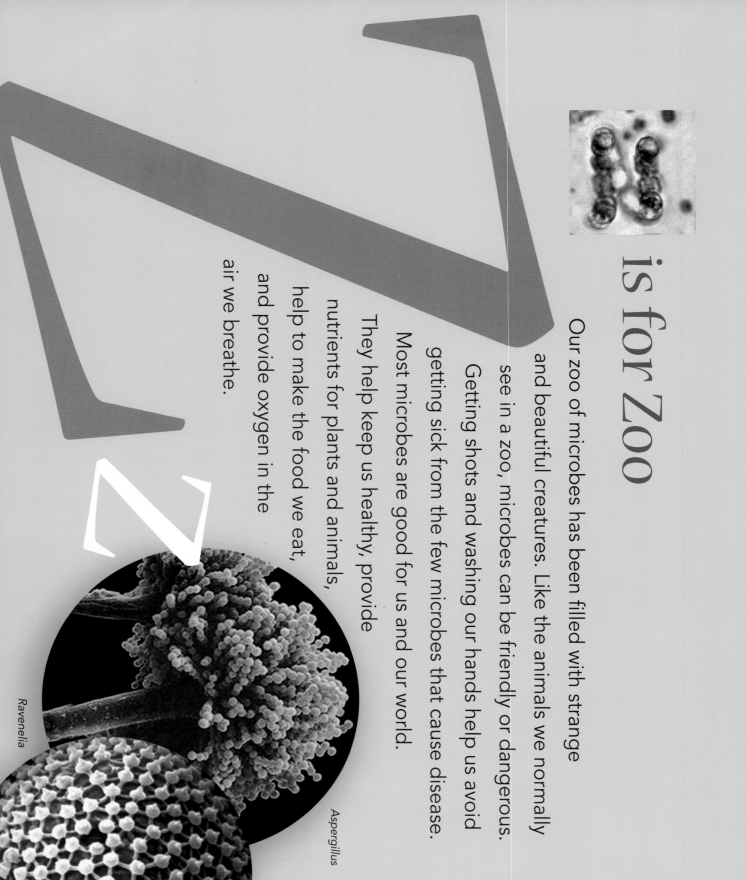

Our zoo of microbes has been filled with strange and beautiful creatures. Like the animals we normally see in a zoo, microbes can be friendly or dangerous. Getting shots and washing our hands help us avoid getting sick from the few microbes that cause disease.

Most microbes are good for us and our world. They help keep us healthy, provide nutrients for plants and animals, help to make the food we eat, and provide oxygen in the air we breathe.

Ravenelia

Aspergillus

Diatom fan on
red algae

Amoeba eating
Staurastrum

Xanthidium

Macrasterias

Escherichia coli

How is the
microbial world
similar to the world
we see every day?
How is it different?

Ditylum

Glossary

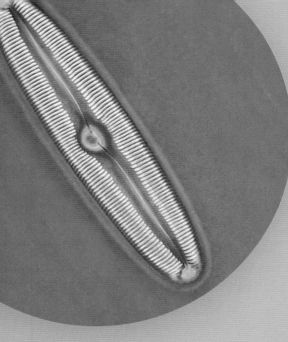

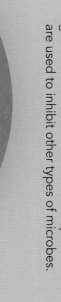

Alcohol (al-ko-hall). Alcohol is an odorless, colorless liquid that evaporates quickly. Microbes break down sugars to form the alcohol found in beer and wines.

Amoeba (a-mee-bah); or Amoebae (a-mee-bee) when referring to more than one organism. An amoeba is a type of microorganism called a protozoan. It moves by oozing portions of itself out and then pulling them back in again.

Antibiotic (an-tee-by-ah-tik). An antibiotic is a medication that kills bacteria or slows their growth. Antibiotics do not affect other microorganisms such as viruses. Other compounds are used to inhibit other types of microbes.

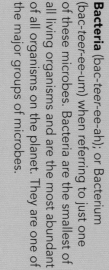

Bacteria (bac-teer-ee-ah); or Bacterium (bac-teer-ee-um) when referring to just one of these microbes. Bacteria are the smallest of all living organisms and are the most abundant of all organisms on the planet. They are one of the major groups of microbes.

Bioremediation (by-oh-re-mee-dee-ay-shun). A process that uses microorganisms to remove pollution from the environment.

Black smoker. Black smokers are cracks in the ocean floor that release heat from under the earth's crust into the water. The very hot water is so rich in minerals it turns black. Some bacteria have adapted to live in this super-hot water and use the minerals as a source of energy.

Carbon dioxide (car-bon dy-ox-ide). Carbon dioxide is a gas that is released by many living organisms and are the most abundant used by plants and some microbes to make the food they need.

Cell. A cell is the smallest unit of life. All living things are made of one or more cells.

Cellulose (sell-you-loze). Cellulose is the sturdy material found in the cell walls of green plants.

Chicken pox. A common disease of childhood, caused by a virus, that causes fever and a bumpy, itchy rash.

Chocolate (yum). Chocolate is made from cocoa beans, which grow tightly packed in a tough husk. Bacteria are used to rot the husk away from the beans; at the same time, they make the beans sweeter. The beans are then roasted, ground fine, and mixed with milk and sugar to make chocolate.

L

Larva (*lar*-vah); or Larvae (*lar*-vee) when there are more than one. An early stage of development in some animals. A larva looks very different from the adult.

Lichen (*lye*-ken). Lichens are made up of two different organisms, microscopic algae (or algae-like bacteria) and a string-like fungus, that live and work together to survive. The fungus part absorbs water and minerals from the environment, and the algae use them to make food for the fungus and itself.

G

Geyser (*guy*-zer). A geyser is a spring that sprays boiling hot water into the air at different times.

Glacier (*glay*-shur). A glacier is a river of ice that slowly flows downhill.

I

Immunity (i-*myoon*-i-tee). The ability of the body to provide long-term protection against infections and toxins produced by microbes.

Influenza (in-floo-*enz*-ah). An illness caused by the influenza virus. Influenza causes a fever, sore throat, muscle aches, a headache, and cough; the person feels tired and sick. The term "flu" sometimes refers to influenza and some-times is used to describe a milder disease not caused by the influenza virus.

D

Dead Sea. A body of water located in the Middle East, at the lowest land point on the planet. It is so salty that only microbes can live in the water.

Diatom (*dy*-ah-tom). Diatoms belong to the group of microbes called algae. They are encased in a two-part wall of silica (*sill*-i-ka), a material that looks like glass. They come in many beautiful shapes and are very common in oceans and fresh water.

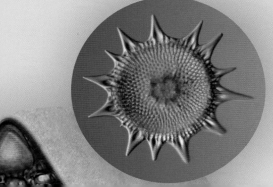

F

Fungi (*fun*-jye); just one is called a Fungus (*fung*-gus). Fungi are organisms like mushrooms that break down dead plant and animal material and recycle the nutrients back into the environment. There are two types of microscopic fungi: molds and yeasts.

M

Measles (mee-zuls). Measles is an illness that includes a fever and a flat red rash. It's caused by a virus. People can get a vaccination to prevent measles.

Microbiologist (my-kro-bye-ol-oh-jist). A scientist who studies microorganisms.

Microorganism (my-kro-or-gan-izm) or Microbe (my-krobe). From "micro" (tiny) and "organism," a microorganism is a living thing that's too small to see. Kinds of microbes include algae, bacteria, yeast, molds, and protozoa, and infectious agents like viruses.

Microscope (my-kro-skope). A microscope is a scientific tool used for viewing objects that are too small to be seen otherwise. "Microscopic" (my-kro-skop-ic) describes something you can only see with a microscope.

Mold. A type of fungus that is composed of microscopic string-like structures that grow in a fuzzy mass as they break down dead plant and animal material. Molds grow in damp areas and are sometimes used to make medicines and foods.

Mumps. Mumps is a common childhood disease that causes a fever and a painful swelling of the glands on the sides of the throat. There is a vaccination to prevent mumps.

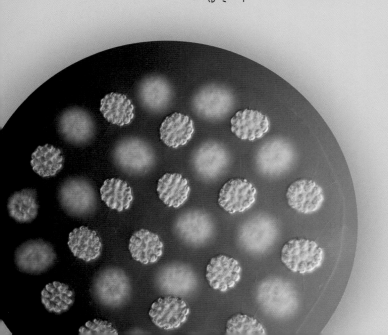

N

Nutrient (new-tree-ent). A nutrient is any substance used by an organism for growth. Some nutrients provide energy for the organism, and others support the organism's ability to grow.

O

Organism (or-gan-izm). A living thing that can grow and reproduce itself. Whales, microbes, reptiles, birds, plants, and kids are all organisms.

Oxygen (ox-i-jen). Oxygen is a gas that many organisms must have to survive. When we breathe, we take in the oxygen we need. Oxygen is made mostly by algae and bacteria in the ocean. Trees and other plants also produce oxygen.

P

Plankton (plank-tun). Plankton are microscopic organisms that drift around with the flow of water. They are the beginning of the food chain for all the animals that live in the water.

Protozoa (pro-to-zo-ah); just one is a Protozoan (pro-to-zo-un). Protozoa are microscopic organisms that cannot make their food from sunlight. They either eat, engulf, or absorb their food. They live in water and moist soils.

Sewage (soo-age). Sewage is water that contains feces and urine and other dirty stuff. It is carried away by sewer pipes and treated to make it safe before it's released back into the environment.

Smallpox (small-poks). Smallpox was a disease caused by a virus that produced a rash of large, pus-filled blisters. Many people died from smallpox, and those who lived were left scarred and sometimes blind. Since smallpox was only found in people and everyone was vaccinated against the disease, smallpox is no longer present anywhere in the world.

T

Termites (ter-mites). A termite looks like a large ant. They can eat wood as a food source because they have microorganisms that help them break the wood down into nutrients.

V

Vaccination (vack-si-nay-shun). Vaccination is a process where a non-disease-causing microbe or part of a microbe is given to a person so the person can develop immunity to a disease.

Virus (vy-rus). A virus is a disease-causing microbe that is smaller than a cell. Viruses enter living cells and cause them to make more viruses.

Vitamins (vy-ta-mins). A vitamin is a substance that is only needed in small amounts but is still required for an organism to live.

W

Warts. A wart is generally a small, rough, raised growth that is caused by a virus.

Whales. Whales are the largest animals on the planet. They live in the oceans. They are not related to fish or sharks, but are mammals that breathe air and provide milk for their young like humans do.

Whooping cough. Whooping cough is a disease caused by bacteria. It causes fever, sore throat, and uncontrollable coughing fits that last so long that a person has trouble getting enough air to breathe. It can cause death in small children but is easily prevented by childhood vaccinations.

Y

Yeast (yeest). Yeasts are single-celled fungi. They can be used to make bread and alcoholic beverages and can also cause disease.

Photo Credits

Sources:

Agriculture Research Service of the United States Department of Agriculture (USDA-ARS); ARS Image Library

American Society for Microbiology (ASM) Archives

Bea Anderson

Rod Anderson (R.A.)

ASM MicrobeLibrary

California Academy of Sciences, Department of Invertebrate Zoology and Geology, Diatom Collection (CAS)

Centers for Disease Control and Prevention (CDC), Public Health Service Image Library (PHIL)

Cushman Foundation for Foraminiferal Research, Inc.

Andrew Davidhazy, School of Photographic Arts and Sciences, Rochester Institute of Technology, Rochester, N.Y.

Donald Ferry

Getty Images

Greenpeace

Dr. Robert Kistler, Bethel University, St. Paul, Minn.

Copyright Dennis Kunkel Microscopy, Inc.

Stephen Nagy

National Aeronautics and Space Administration (NASA)

National Library of Medicine, National Institutes of Health (NLM/NIH)

National Oceanic and Atmospheric Administration (NOAA)

National Science Foundation (NSF)

Debra Naylor

Ohio Department of Agriculture

D. J. Patterson, provided courtesy of the micro*scope website at http://micro*scope.mbl.edu

Photodisc

John C. Priscu, Montana State University, Bozeman

Josh Rubinstein

Robert R. Tupper, Jr.

U.S. Department of Energy (USDOE)

U.S. Environmental Protection Agency (USEPA)

U.S. Fish and Wildlife Service (USFWS)

U.S. National Park Service (USNPS)

Wim van Edmond, Micropolitan Website

Morgan Vis, Ohio University Alage image database

Denise Weitrich

Petr Znachor

(All other images from iStockphoto)

Introduction:

Florida swamp, Donald Ferry; Lake Mitchell, South Dakota, R.A.; Closterium and miscellaneous algae, Petr Znachor; Cymbellan, Petr Znachor; Zygnema, R.A.; Green algae, Petr Znachor; Pleodorina indica, Petr Znachor; Staurastrum subtelliferum and Bambusina brebissoni, Wim van Edmond

Bacteria (background), NASA; Gracie playing soccer, Denise Weitrich; Big horn sheep, Custer State Park, South Dakota, R.A.; Microscope, Photodisc; Decaying tree, Bea Anderson; Staphylococcus aureus, Beltsville Agricultural Research Center Plant Sciences Institute, USDA-ARS; Spirillum volutans, R.A.; Leptospira, Janice Carr, CDC PHIL; Vibrio vulnificus, Janice Carr, CDC PHIL

Donkey, R.A.; Bison, Photodisc; Cows grazing, Bruce Fritz, USDA-ARS; Technician Melissa Goff and animal scientist Kathy Soder studying stuff from a cow's stomach, Stephen Ausmus, USDA-ARS; Rumen bacteria, D. J. Patterson; Rumen bacteria, D. J. Patterson

Diatom rosette (background), Stephen Nagy; Ditylum, NOAA Center for Coastal Environmental Health and Biomolecular Research; Pinnularia, Stephen Nagy; Lake North Cascades National Park, USNPS Digital Image Archives; Fossil diatoms from Barbados, Wim van Edmond; Marine diatom fossil, Stephen Nagy; Actinoptychus helio-pelta, Stephen Nagy; Sunset near Cancun, R.A.

Glacier in the Wright Valley, Antarctica, Peter Doran, NSF; Black smoker, P. Rona, OAR/National Undersea Research Program, NOAA; Storm clouds, Photodisc; Coscinodiscus, Wim van Edmond; Achnanthes, Wim van Edmond; Periphaena decora, Wim van Edmond; Licmophora, Wim van Edmond

Microscopic view of a lichen, D. J. Patterson; Lichen on tree, Boreal Forest, Josh Rubinstein; Red lichens on rocks, Lamar Valley, Yellowstone National Park, J. J. Schmidt, USNPS

Bacteria (background), CDC PHIL; Propionibacterium, Copyright Dennis Kunkel Microscopy, Inc.; Cocoa beans, Keith Weller, USDA-ARS Photo Library; Lactobacillus, Eric Johnson and Byron Brehm-Stecher, ASM MicrobeLibrary; Yeast cells, D. J. Patterson

Bacteria (background), CDC PHIL; Enterococcus, Copyright Dennis Kunkel Microscopy, Inc.; Three types of bacteria, Copyright Dennis Kunkel Microscopy, Inc.; Escherichia coli, Eric Erbe, digital colorization by Christopher Pooley, USDA-ARS; Beans, Keith Weller, USDA-ARS

Lactobacillus, Copyright Dennis Kunkel Microscopy, Inc.; Staphylococcus epidermidis, Copyright Dennis Kunkel Microscopy, Inc.; Vitamins, Photodisc; Staphylococcus aureus on human hair, Copyright Dennis Kunkel Microscopy, Inc.; Freewheeling, Getty Images

Measles virus, CDC PHIL; Red blood cells and bacteria-eating cells, Copyright Dennis Kunkel Microscopy, Inc.; Child receiving a vaccination, CDC PHIL; Bacteria-eating blood cell, Copyright Dennis Kunkel Microscopy, Inc.; "Rubella Fighter": child with a vaccination card, CDC PHIL

Cowpox lesions on the arm of Sarah Nelmes, NLM/NIH; Smallpox virus, CDC PHIL; Edward Jenner, NLM/NIH; The Cow Pock-or-The Wonderful Effects of the New Inoculation! Editorial cartoon by British satirist James Gillray 1802, NLM/NIH; Modern vaccination, Barbara Rice, CDC PHIL

Diatoms arranged in a rosette, Stephen Nagy; Nauplii, NOAA Center for Coastal Environmental Health and Biomolecular Research; Arranged diatoms, Stephen Nagy; Globigerina, D. J. Patterson; Bacillaria paxillifer, Wim van Edmond; Kaleidoscope, Bea Anderson

Arthroderma, Lucille K. George, CDC PHIL; *Mucor,* Lucille K. George, CDC PHIL; Moldy grapefruit, Scott Bauer, USDA-ARS; *Hypocrea,* USDA-ARS; Mold colonies in petri dishes, Scott Bauer, USDA-ARS; Moldy wall, USEPA; *Botrytis,* Lucille K. George, CDC PHIL; Antibiotic pills, R.A.

Termites, USDA-ARS; Researchers inspecting termite damage, USDA-ARS; *Trichonympha,* Wim van Edmond; *Snyderella,* D. J. Patterson

Fall colored leaf in stream, Yellowstone National Park, J. Schmidt, ©1977 USNPS; Aspen woods, Elk Creek, Yellowstone National Park, William W. Dunmire, ©1969 USNPS; *Pseudomonas,* Copyright Dennis Kunkel Microscopy, Inc.; *Nitrosomonas,* Copyright Dennis Kunkel Microscopy, Inc.

Man pumping gas that contains ethanol from corn, Ohio Department of Agriculture; Gold, Photodisc; Food technologist Gul Uludogan making cheese, Keith Weller, USDA-ARS; Antibiotics, R.A.; Oiled seabird, *Braer* oilspill, Shetland Islands, Scotland, Greenpeace/Roger Grace

Grass and trees at Thumb Creek, Yellowstone National Park, R. G. Johnsson, USNPS; Global chlorophyll production map, NASA; Lead tree, *Leucaena leucocephala,* Scott Bauer, USDA-ARS Image Library; *Eudorina elegans,* Wim van Edmond; *Pediastrum* and *Coelastrum,* Petr Znachor; *Stentor* with the endosymbiotic algae *Chlorella,* Wim van Edmond; *Euglena,* Wim van Edmond

Rotavirus, Erskine Palmer, CDC PHIL; Avian influenza A H5N1 virus, C. Goldsmith, J. Katz, and S. R. Zaki, CDC PHIL; Droplet spray from a sneeze, Andrew Davidhazy; Microbes growing from a sneeze, ASM Archives; Ebola virus, Frederick A. Murphy, CDC PHIL

Amoeba, Donald Ferry; Ciliate, Donald Ferry; *Stylonychia,* Wim van Edmond; *Vortecella,* Wim van Edmond; *Diophrys,* Wim van Edmond; White tail deer, Photodisc; Wolf, Photodisc

Gray whale, Camille Goebel, Alaska Fisheries Science Center, National Marine Mammal Laboratory, NOAA; *Achnanthes,* Wim van Edmond; Calanoid copepod, Wim van Edmond; Copepod, Wim van Edmond; *Noctiluca scintillans,* Wim van Edmond; *Lithodesmium intricatum,* Wim van Edmond; Starfish larvae, Wim van Edmond

Plant pathologist Rick Bennett examines fungi that may be used for biological control of weeds, Scott Bauer, USDA-ARS; *Ophiocytium,* D. J. Patterson; *Gallonella,* D. J. Patterson; Blood with cells that eat microbes, Copyright Dennis Kunkel Microscopy, Inc.; *Anabaena,* Petr Znachor

Black smoker, NASA; *Bacillus infernos,* Subsurface Microbial Culture Collection, USDOE; Microbial streamers, Yellowstone National Park, Debra Naylor; Isanotski Peak Volcano, Unimak Island, Alaska, John Sarvis, USFWS; Nostoc in glacial ice, John C. Priscu, Montana State University, Bozeman; *Halobacterium,* NASA

Roots with nodules, USDA-ARS; *Bradyrhizobium,* Copyright Dennis Kunkel Microscopy, Inc.

Brewer's yeast, Robert R. Tupper, Jr.; Yeast cell (*Saccharomyces cerevisiae* showing bud scars), Alan Wheals and Anna Cosney, ASM MicrobeLibrary

Sewage treatment, R.A.; *Zoogloea,* D. J. Patterson; Protozoan, Donald Ferry; Ciliate, Donald Ferry; Amoeba, Donald Ferry; Ciliate, Donald Ferry; Sewage treatment plant, Photodisc

Aspergillus, Copyright Dennis Kunkel Microscopy, Inc.; *Ravenelia spinulosa,* J. R. Hernandez et al., USDA-ARS; *Ditylum,* Wim van Edmond; *Escherichia coli,* Copyright Dennis Kunkel Microscopy, Inc.; *Macrasterias rotata* cell division, Wim van Edmond; Fan-shaped diatom on red algae, Wim van Edmond; *Amoeba proteus* eating *Staurastrum,* Wim van Edmond; *Xanthidium,* Petr Znachor

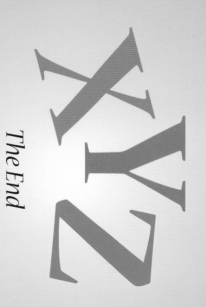

The End

How big is small?

If a virus were the size of a grain of sand...

Then a bacterium would be the size of a grape...

The head of a pin would be the size of a whale...